SIMPLE SCIENCE

FUN

SIMPLE SCIENCE

FUN

by
WENDY EVANS

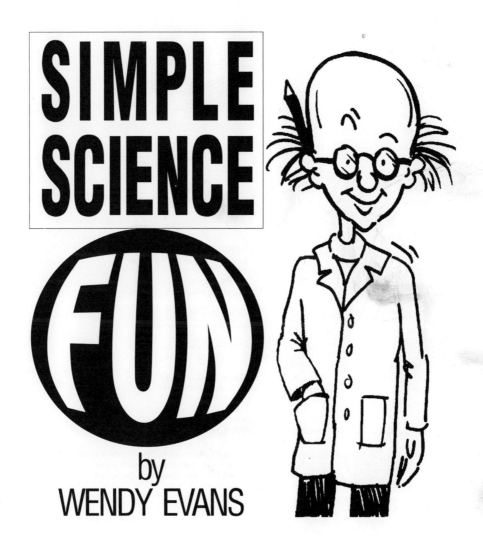

ASHTON SCHOLASTIC
AUCKLAND SYDNEY NEW YORK TORONTO LONDON

Published by Ashton Scholastic, 1994

Ashton Scholastic Ltd
Private Bag 92801, Auckland, New Zealand.

Ashton Scholastic Pty Ltd
PO Box 579, Gosford, NSW 2250, Australia.

Scholastic Inc
555 Broadway, New York, NY 10012, USA.

Scholastic Canada Ltd
123 Newkirk Road, Richmond Hill, Ontario L4C 3G5, Canada.

Scholastic Publications Ltd
7-9 Pratt Street, London, NW1 0AE, England.

Copyright © Wendy Evans, 1994
ISBN 1 86943 169 3

9 8 7 6 5 4 3 2 1   4 5 6 7 8 9 / 9

Edited by Penny Springthorpe
Designed by Christine Dale
Illustrations by Jenny Jeffries
Typeset in 12/16pt Southern
Printed in New Zealand by BPG Ltd

Contents

Balloon on a Bottle

You will need:
- a balloon
- a 1-litre plastic bottle
- 2 medium-sized pots
- boiling water
- cold water
- a tray of ice cubes

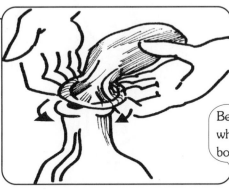

Be very careful when using the boiling water.

1   Stretch the balloon over the top of the bottle.

2   Half fill one pot with boiling water, and half fill the other with cold water and ice cubes.

3   Stand the bottle in the boiling water for 30 seconds. Watch the balloon.

4   Now stand the bottle in the ice cold water for 30 seconds. Watch the balloon.

Simple Science

When the air is heated it expands (takes up more space) causing the balloon to inflate. When the air is cooled, it contracts (takes up less space) causing the balloon to shrink into the bottle.

Balloon Magic

You will need
- a long balloon
- tiny pieces of paper

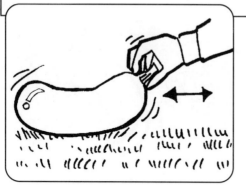

1 Blow up the balloon and tie a knot in the end. Rub it back and forth quickly on the carpet.

2 Bring the balloon towards the tiny pieces of paper.

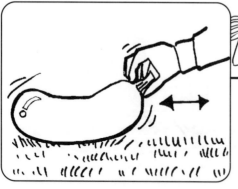

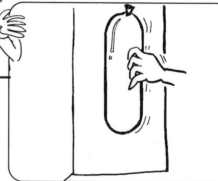

3 Again rub the balloon quickly on the carpet.

4 Now hold it up against the wall and let go. Magic!

Simple Science

The surface of the balloon has become charged with static electricity, and this is what attracts the pieces of paper.

Bottle Rocket

You will need:
- a large plastic bottle with cap
- scissors
- 2 straws (about 10 cm long)
- Sellotape
- Plasticine

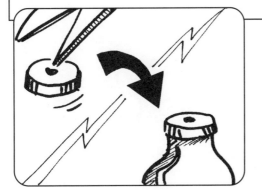

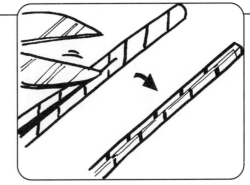

1   Using the point of the scissors, pierce a small hole in the centre of the cap and screw it firmly onto the bottle.

2   Slit one of the straws in half lengthways. Roll it up tightly and tape down the seam to seal it. You now have a much thinner straw.

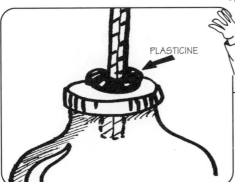

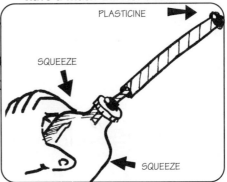

3   Slide 2 cm of the straw through the hole in the cap and seal around the hole with Plasticine.

4   Seal the top of the second straw with Plasticine and slide it over the thinner straw.
    Aim safely, and squeeze the bottle hard.

Simple Science

When the bottle is squeezed hard, the air inside is forced out through the narrow straw. The pressure of this air propels the rocket away from the bottle.

Bubble Pictures

1 Mix 2 teaspoons of food colouring and 1 teaspoon of detergent together and add to $1/4$ cup of water.

2 Slide the end of the straw into the liquid and blow gently until bubbles rise over the top of the cup.

3 Place the paper over the bubbles to make a print. To make more prints, blow more bubbles.

4 For variation, you can set up a number of cups using different colours.

Simple Science

The bubbles are formed by forcing air into the detergent and water. They are always spherical (round), because the outside of the bubble is under the least amount of stress in this shape.

Candle in a Jar

You will need:
- a candle (about 8 cm long)
- an old breakfast bowl
- water
- matches or lighter
- a clear jar

You may need help to stick the candle to the bowl.

1. Stick the candle to the bottom of the bowl with melted wax.

2. Carefully pour $1/2$ cup of water into the bowl and light the candle.

3. Turn the jar upside down and cover the candle.

4. Watch the water level and the candle flame.

Simple Science

Oxygen is a gas which makes up about one-fifth of the air we breathe. The flame burns the oxygen, reducing the amount of air in the jar. The water then rises into the space that was taken up by the oxygen.

5

Diver in a Bottle

You will need :
- a large plastic bottle, with cap
- water
- a pen top
- Plasticine

1 Fill the plastic bottle to the top with water.

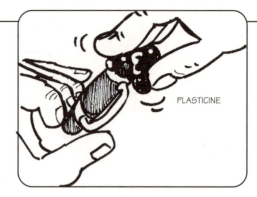

PLASTICINE

2 Seal the open end of the pen top with Plasticine.

3 Drop the pen top into the water and screw on the cap tightly.

4 Squeeze the bottle with your hands. Now let go.

Simple Science

Squeezing the bottle causes the air trapped inside the pen top to be compressed (squashed). This reduces its buoyancy (ability to float), making it sink. When the bottle is released the air expands again, increasing the buoyancy of the pen top and making it rise to the top.

Egg in a Bottle

You will need:
- a hard-boiled, shelled egg
- a piece of paper towel (about 5 cm square)
- matches or lighter
- a wide-necked glass bottle
- a cup of cold water

You may need an adult to watch you light the paper towel.

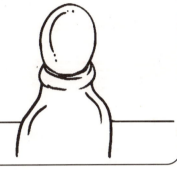

1 Screw up and light the piece of paper towel, and drop it burning into the bottle.

2 Quickly dip the egg in the cold water and sit it in the neck of the bottle, point downwards.

3 Watch the egg very carefully.

Simple Science

When the air inside the bottle is heated it expands (takes up more space). When the flame goes out and the air cools down again, it contracts (shrinks), sucking the egg into the bottle.

Ice Bowl

You will need

- 2 bowls (one slightly smaller than the other)
- water
- a paperweight or small rock
- flower petals and leaves

1 Sit the smaller bowl inside the larger one and carefully pour water in between the two.

2 When there is approximately 2 cm of water beneath the small bowl, place the weight inside it to stop it floating any higher.

3 Push petals and leaves down into the water between the bowls. Freeze overnight.

4 To free the ice bowl, first pour lukewarm water into the small bowl and gently remove it. Then stand the big bowl in lukewarm water and lift out the ice bowl.

Simple Science

When water is cooled to 0° centigrade, it freezes and becomes solid (ice). When the ice is warmed, it becomes liquid again. When water is heated to about 100° centigrade, it becomes a gas (steam).

Magic Cardboard

You will need:
- a clear drinking glass
- water
- a cardboard square, large enough to cover the top of the glass

1   Fill the glass with water.

2   Place the cardboard square over the top.

3   Hold the cardboard firmly in place with one hand and, holding the glass over the sink, turn it upside down.

4   Slowly remove your hand from the cardboard.

Simple Science

The water inside the glass forms a seal between the cardboard and the rim. Air presses evenly around the outside of the glass, keeping the cardboard in place.

Magic Ink

You will need

- white paper
- a large art paintbrush
- 1 dessertspoon food colouring (in a cup)
- a cotton bud
- 1 dessertspoon household bleach (in an old cup)

Bleach is poisonous. Take care not to get it on your skin or clothing.

1   Use the paintbrush to paint the paper all over with the food colouring. Leave to dry.

2   Dip one end of the cotton bud into the bleach.

3   Use the cotton bud to draw a picture on the coloured paper, redipping in the bleach as necessary.

4   Leave to dry on a flat surface. Observe what happens.

Simple Science

Bleach is a chemical which strips away colour.

Magnifying Glass

You will need
- small objects (coin, screw, paperclip)
- a medium-sized, clear bowl
- cling film
- a large rubber band

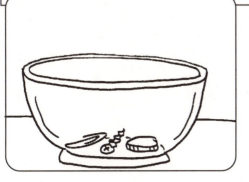

1   Place the small objects in the bowl.

2   Cover the top of the bowl with cling film and push down gently to make a hollow.

3   Secure the cling film with the rubber band.

4   Fill the cling film hollow with water, then look at the objects through the top and sides of the bowl.

Simple Science

When light passes through water it is bent slightly. Together, the bent light and the curved cling film have a magnifying effect, making the objects look bigger than they really are.

11

Mouldy Bread

You will need

- a slice of bread (wholemeal works best)
- warm water
- a plastic bag
- a wire tie
- a magnifying glass

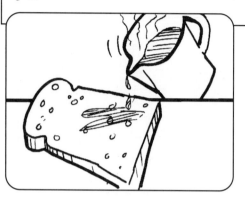

1   Dampen the bread with warm water.

2   Place the bread in the plastic bag and close with the wire tie.

3   Leave in a dark place.

4   Using the magnifying glass, examine the bread through the plastic bag every day.

Simple Science

Fungi spores float in the air around us. When they encounter a favourable surface, such as damp bread, they begin to grow and multiply.

Photosynthesis

You will need
- a leaf (still attached to a growing plant)
- a small strip of paper
- 2 paperclips

1   Partly cover the leaf with the paper strip.

2   Hold in place with the paperclips.

3   Leave for a few days, then remove paperclips and paper.

4   Compare it to the other leaves on the plant.

Simple Science

The green part of a leaf traps sunlight and uses it, together with water and air, to make food for the plant. This is called photosynthesis. By covering up part of the leaf, you have interrupted this process.

Potato Slices

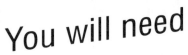
You will need

- 2 shallow bowls
- salt

- 2 slices of raw potato (the same size)

1   Fill one bowl with plain water.

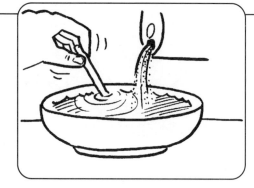

2   Fill the other bowl with very salty water.

3   Place one potato slice in each bowl and leave to stand for several hours.

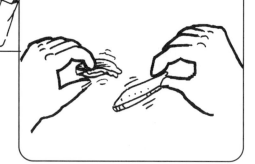

4   Take the slices out and compare them.

Simple Science

The cell walls of the potato allow plain water to pass in and out freely, but do not allow salty water in.   The potato slice in the salty water shrinks because liquid leaving the potato cannot be replaced.

Psychedelic Milk

You will need :
- milk
- a breakfast bowl
- food colouring (4 different colours)
- detergent
- an eye dropper or straw

To use the straw as an eye dropper, place one end in the liquid, seal the other end with one finger and lift the straw. Lift your finger momentarily to allow a drop to fall.

1   Pour some milk into the breakfast bowl.

2   Using the eye dropper or straw, drop a little food colouring into the centre of the milk.

3   Add 1 or 2 drops of detergent.

4   Watch carefully for several minutes.

Repeat steps 2 and 3 using different colours.

Simple Science

The surface of the milk has a sort of skin, known as surface tension. The detergent breaks this skin, causing the milk to move around. Adding colour allows us to see this happening.

Rubber Egg

You will need
- an egg
- a small bowl
- vinegar
- a spoon
- a large bowl

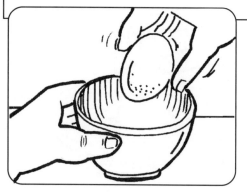

1 Place the egg in the small bowl.

2 Pour in enough vinegar to cover the egg and leave to stand for two days.

3 Use the spoon to remove the egg from the vinegar.

4 Bounce the egg carefully in the large bowl.

Simple Science
The chemicals in the vinegar act on the eggshell, changing its texture. This is called a chemical reaction.

Simple Kaleidoscope

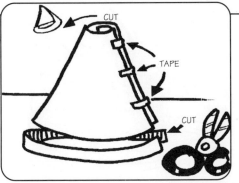

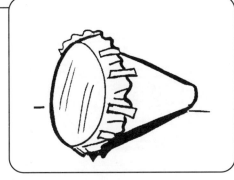

1 Make the cardboard into a cone and tape in place. Cut a small piece off the narrow end and trim the cardboard at the wide end so it sits flat.

2 Cover the wide end with yellow Cellophane and tape in place.

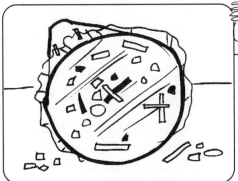

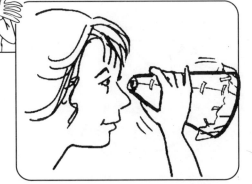

3 Glue the small pieces of Cellophane on to the yellow Cellophane.

4 Leave to dry overnight. Hold the narrow end to your eye and turn the cone slowly.

Simple Science

Red, yellow and blue are primary colours. By combining them, new colours are created: red + yellow = orange, red + blue = purple, blue + yellow = green.

Sinking Cork

You will need
- a clear bowl
- a cork
- a clear drinking glass

1 Half fill the bowl with tap water and float the cork in it.

2 Try to sink the cork with your fingers.

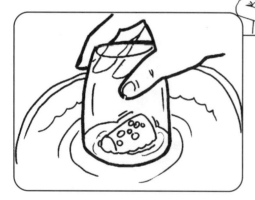

3 Now hold the glass upside down over the cork.

4 Firmly push the glass to the bottom of the bowl. Watch the cork.

Simple Science

The air inside the glass is compressed (squashed), making the pressure stronger than the air pressure on top of the water in the bowl. This stronger pressure forces the water out of the glass. The cork then drops to the bottom of the bowl through the pull of gravity.

Snow Scene

You will need:
- a clear plastic jar with lid
- Plasticine
- a small plastic toy figure
- ¹/₂ cup dessicated coconut or silver glitter
- masking tape or Sellotape

1 Stick a piece of Plasticine to the inside centre of the lid.

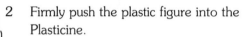

2 Firmly push the plastic figure into the Plasticine.

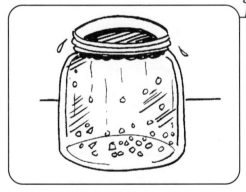

3 Fill up the jar with tap water and add the coconut or glitter.

TAPE

4 Screw on the lid tightly and seal with tape. Turn the jar upside down and shake well.

Simple Science

Coconut and glitter are heavier than water, so they sink when the jar is still. (Some coconut may dissolve in the water over time.)

Soap Boat

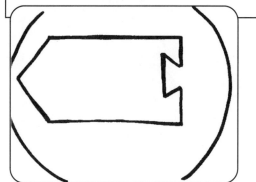

1 With a felt pen, draw the boat shape on the plastic lid and cut out.

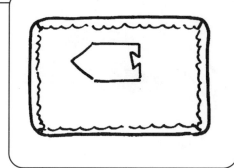

2 Float the boat in the sink or bath.

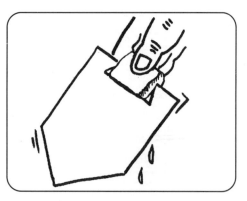

3 Remove the boat and squash the soap onto the notch on the back of the boat.

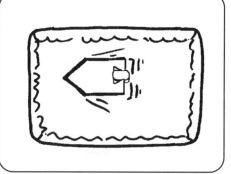

4 Float the boat again and see what happens.

Simple Science

The soap breaks the surface tension of the water (see Psychedelic Milk) at the back of the boat, propelling it towards the part of the water where the surface tension is still strong.

Spiky Caterpillar

You will need:
- 60 gm packet of grass seed
- 10 cups soil or potting mix
- a large bowl
- leg from a pair of old pantyhose
- 2 buttons for eyes
- needle and thread

Ask someone to hold the pantyhose leg while you fill it.

1   Mix grass seed and soil together in the bowl.

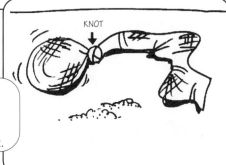

KNOT

2   Pour two cups of soil into the pantyhose. Tie a knot immediately above the soil.

3   Repeat Step 2 until the pantyhose leg is full (about five knots). Sew eyes in place.

4   Place your caterpillar in the bowl in a sunny place, and water daily.

Simple Science

When seeds are placed in an ideal environment, they will grow very well. They need good soil, water and sunlight.

Spinning Eggs

You will need
- 2 eggs (1 raw, 1 hard-boiled)
- a felt pen

1   Mark the hard-boiled egg with the felt pen so you know which one is cooked.

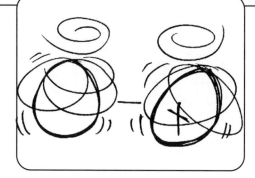

2   Spin the two eggs on a flat surface as fast as you can.

3   Put your fingers on the eggs for a moment, then quickly let go.

4   Watch the two eggs carefully.

Simple Science

When the raw egg is spun quickly, then stopped, the fluid inside the shell keeps going, causing the egg to move again. This is called inertia. The cooked egg doesn't move because the inside is solid.

22

Suspended Egg

You will need
- a clear container
- a fresh egg
- a spoon
- salt

1 Fill container with cold water and lower in the egg on the spoon.

2 Remove the spoon, note the position of the egg, then take it out again.

3 Add salt to the water and stir. Keep adding and stirring until no more salt can dissolve.

4 Carefully lower the egg back into the salt water solution.

Simple Science

The water and salt are 'in solution', which means the water has dissolved all the salt it can hold. This solution is more dense (heavier) than the egg, which is why the egg floats.

Talking Cans

You will need

- 2 clean, empty cans
- a hammer
- a nail
- 2 toothpicks or matchsticks
- string, 3-4 metres long

1 Use the hammer and nail to make one hole in the centre bottom of each can.

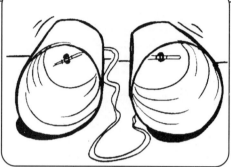

2 Thread the string through the holes, and tie on the toothpicks or matchsticks so they rest inside the bottom of each can.

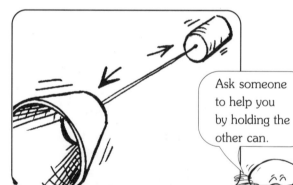

3 Pull the cans apart until the string is stretched taut.

Ask someone to help you by holding the other can.

HELLO
HELLO

4 Take turns talking into one can and listening through the other.

Simple Science

The sound waves from your voice travel along the string as vibrations. They are then changed back into sound waves by the can at the other end of the string.

Water Bottles

1 Half fill the three bottles with cold water.

2 Pour the glitter or coconut into the first bottle. Pour the oil into the second bottle. Push polystyrene chips into the third bottle.

3 Screw the caps on the bottles and seal with tape.

4 Shake each bottle, and compare what happens inside each one.

Simple Science

The glitter or coconut sinks, because it is more dense (heavier) than water. The oil seems to mix with the water at first, but it will gradually separate and float to the top of the water because it is less dense (lighter). Polystyrene chips float because they are extremely light and non-absorbent (don't soak up any water).